TULSA CITY-COUNTY LIBRARY

For Emalia, Charlie, Jackson, Vivian,
and Remy, with lots of love —G.T.

For Molly, my hairy friend —A.J.

Text copyright © 1979 by Graham Tether
Illustrations copyright © 2019 by Andrew Joyner

All rights reserved.
Published in the United States by Random House Children's Books,
a division of Penguin Random House LLC, New York.
Originally published in different form by
Random House Children's Books, New York, in 1979.

Random House and the colophon and Bright and Early Books and the
colophon are registered trademarks of Penguin Random House LLC.
The Cat in the Hat logo ® and © Dr. Seuss Enterprises, L.P. 1957,
renewed 1986. All rights reserved.

Visit us on the Web!
rhcbooks.com

Educators and librarians, for a variety of teaching tools, visit us at
RHTeachersLibrarians.com

Library of Congress Cataloging-in-Publication Data
Names: Tether, Graham, author. | Joyner, Andrew
(Illustrator), illustrator.
Title: The hair book / by Graham Tether ; illustrated by Andrew Joyner.
Description: First edition. | New York : Beginner Books,
a Division of Random House, [2018]. | Series: A bright and early book
Originally published with a different illustrator: New York :
Beginner Books, 1979. | Audience: Ages 2–5.
Identifiers: LCCN 2017020621 (print) | LCCN 2017025830 (ebook)
ISBN 978-1-5247-7340-3 (trade) | ISBN 978-1-5247-7341-0 (lib. bdg.) |
ISBN 978-1-5247-7342-7 (ebook)
Subjects: LCSH: Hair—Juvenile literature.
Classification: LCC QP88.3 (ebook) | LCC QP88.3 .T48 2018 (print) |
DDC 573.5/8—dc23

Printed in the United States of America
10 9 8 7 6 5 4 3 2 1
First Edition

Random House Children's Books supports the First Amendment
and celebrates the right to read.

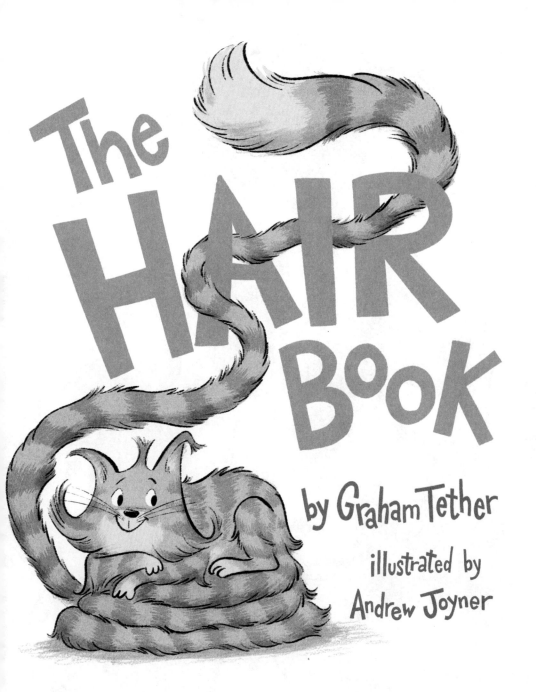

The HAIR BOOK

by Graham Tether

illustrated by Andrew Joyner

A Bright and Early Book
From BEGINNER BOOKS®
A Division of Random House 🏠 New York

Wig hair.

Bear hair.

Everybody seems to wear hair.

So many different kinds of hair!

Lambs have woolly coats to wear.

Pigs have bristles.
They don't care.

And porcupines
have quills that scare.

You could make a ladder with your hair . . .

or lasso broncos at the fair.

you could use your hair
to water-ski!

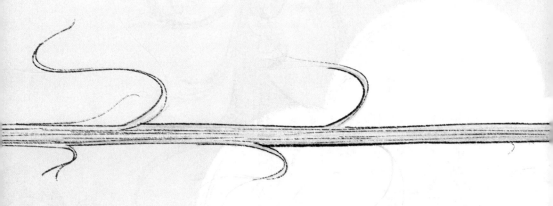

Not one hair.
Not anywhere.